OCEAN MAMMALS

SEA OTTERS

by Julie Murray

Cody Koala

An Imprint of Pop!
popbooksonline.com

Hello! My name is Cody Koala

This book is filled with videos, puzzles, games, and more! Scan the QR codes* while you read, or visit the website below to make this book pop.

popbooksonline.com/sea-otter

*Scanning QR codes requires a web-enabled smart device with a QR code reader app and a camera.

abdobooks.com

Published by Pop!, a division of ABDO, PO Box 398166, Minneapolis, Minnesota 55439.

Printed in the United States of America, North Mankato, Minnesota.

042025
082025

THIS BOOK CONTAINS RECYCLED MATERIALS

Cover Photo: Getty Images
Interior Photos: Shutterstock Images; Getty Images
Editors: Elizabeth Andrews and Grace Hansen
Series Designer: Candice Keimig

Library of Congress Control Number: 2024948402

Publisher's Cataloging-in-Publication Data

Names: Murray, Julie, author.
Title: Sea otters / by Julie Murray
Description: Minneapolis, Minnesota : Pop!, 2026 | Series: Ocean mammals | Includes online resources and index
Identifiers: ISBN 9781098247768 (lib. bdg.) | ISBN 9781098248307 (ebook)
Subjects: LCSH: Sea otter--Juvenile literature. | Otters--Juvenile literature. | Sea otter--Behavior--Juvenile literature. | Marine mammals--Juvenile literature. | Aquatic mammals--Juvenile literature. | Sea animals--Juvenile literature.
Classification: DDC 599.769--dc23

Table of Contents

Chapter 1

Floating Otters

Sea otters are **mammals** that live in the Pacific Ocean. They swim in the cold waters off the coasts of North America and Asia. Most live off the coast of Alaska.

Watch a video here!

Sea otters can grow up to 4 feet (1.2m) long and weigh up to 100 pounds (45kg). They have a wide nose and small eyes and ears.

The sea otter is the smallest **marine** mammal in the world.

Where Sea Otters Live

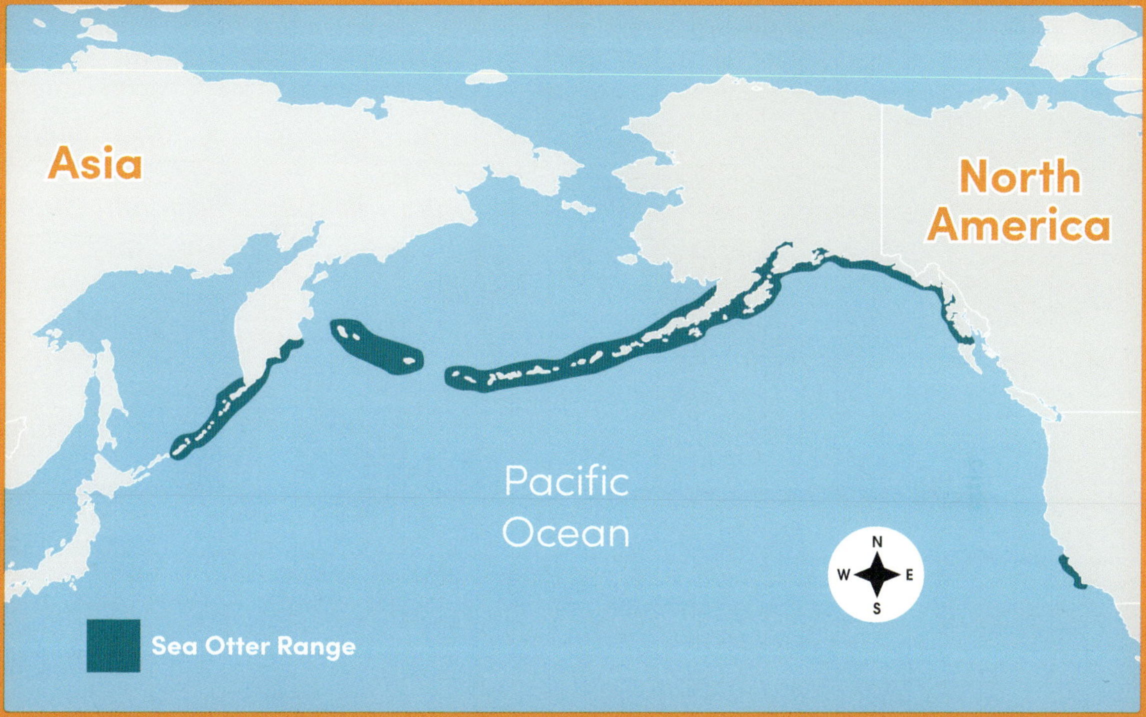

Sea otters have thick, brown fur. Their fur traps air next to their skin to keep them warm. It is also **waterproof**. Sea otters have the thickest fur of all animals!

A sea otter spends a lot of time floating on its back. It does everything from **grooming** and eating to

resting on its back! Sea otters spend much of their day grooming their fur.

Sea otters are excellent swimmers. They can swim over 5 miles per hour (8kph). They have **webbed** back feet to help them swim. A long, flat tail helps them **steer**.

Chapter 2

Kelp Forests

Sea otters are important to ocean life. They keep **kelp** forests healthy. They do this by feeding on **sea urchins**. Sea urchins can destroy kelp forests by eating too much kelp.

Kelp forests provide food, shelter, and oxygen for ocean life.

Learn more here!

Chapter 3

Food

Sea otters can dive 350 feet (107m) to find food. They eat animals such as **sea urchins**, clams, and crabs. They have strong mouths to break open the hard shells.

Sea otters also use rocks to open shells.

Explore links here!

Sea otters bring food to the surface to eat. They have loose skin folds under their arms. They store food

and rocks here. The folds let them carry more things when finding food on the ocean floor.

Chapter 4

One Pup

Females usually give birth to one pup at a time. The pup is 5 pounds (2.3kg) at birth. It has thick fur to keep it warm. The thick fur helps the pup float too.

Sea otters can live for up to 20 years.

Complete an activity here!

Making Connections

Text-to-Self

Sea otters float on their backs. Can you do the back float in the water? If so, how does it feel when you are floating on your back?

Text-to-Text

Have you read any books about other animals that live in the ocean? How are they like sea otters? How are they different?

Text-to-World

Sea otters have thick fur. Can you name other animals that have thick fur too?

Glossary

groom – to clean and make neat in appearance.

kelp – a large, tough, brown seaweed. A kelp forest is an underwater area with lots of kelp.

mammal – a warm-blooded animal that produces milk for its young and is usually covered with hair.

marine – having to do with the sea.

sea urchin – a small sea animal with a soft body inside a hard, round shell covered with spines.

steer – to control the direction of travel.

waterproof – not letting water through.

webbed – when the toes are connected by skin.

Index

Online Resources

popbooksonline.com

Thanks for reading this Cody Koala book!

This book is filled with videos, puzzles, games, and more! Scan the QR codes* while you read, or visit the website below to make this book pop.

popbooksonline.com/sea-otter

*Scanning QR codes requires a web-enabled smart device with a QR code reader app and a camera.